# SECOND TOUCH

# SECOND TOUCH

## Options and Alternatives
## When Facing an Unplanned Pregnancy

Dr. Jack Monaco, MD

Foreword by Jennifer O'Neill

Franklin, Tennessee

ISBN: 979-8-218-63397-4

Cover design by LACreative.
Cover art by ROI Design Group.
Page design by Win-Win Words LLC.

Printed in the United States of America.

# Contents

*This book is dedicated to women around the world who find themselves in the difficult position of having an unplanned pregnancy. It is my fervent desire that, in some way, the words, thoughts, information, and guidance contained in this book will help you see the sanctity of life and choose life over abortion. In that way, I believe that you will be blessed immensely.*

# Foreword

I'VE BEEN KNOWN TO SAY, "You can't be a little bit pregnant." You either are, or you are not. *True?* To that point, if you find yourself with an unexpected or unwanted pregnancy, *Please*, at least make a fully informed decision about your baby's future. After all, it's a matter of life or death.

So, if you, or someone you know, is grappling with the difficult and complex decision of whether or not to have an abortion, my prayer is that you will seriously consider the medical, scientific, and spiritual truths found in this invaluable book by Dr. Jack Monaco. He will answer questions you may have about what is developing in the womb during a pregnancy: He or she is a baby, not a "blob of tissue" or a "cluster of cells" or

"A 'nothing,' especially under three months' gestation."

*That lie* from the pit of hell was what I was told when I was being pressured to abort the pregnancy that I was carrying in 1973.

There were no ultrasounds back then for me to see the humanity of my baby from early on in my pregnancy. FYI: Since the advent of the ultrasound, well over 95 percent of abortion-minded parents *choose life* for their child once they have seen their baby in the womb, as well as have heard its heartbeat!

In my case, my fiancé (the father of my baby) adamantly did *not* want our child: I went to my parents for advice, and they said I could always have another baby later—after the wedding—at a more appropriate time. . . .

My friends pointed out that abortion was *easy* and *legal*. I didn't have my faith then, and I eventually folded under the father's mounting threats and submitted to the "simple procedure"—*abortion*.

Following the death of my baby, I profoundly hated myself for being so weak under adverse pressure and for failing to protect my child. I've regretted for my entire life that uniformed, heart-wrenching decision I made long ago.

The aftermath of abortion not only takes the life of a child, it also profoundly hurts their mothers physically, emotionally. and spiritually. It also destroys families while it corrupts society, in general, along with one's personal moral compass. (For more statistics and information about the full effects of abortion, please read any of my books that address the issue: *Surviving Myself, From Fallen to Forgiven*, or *You're Not Alone*).

Thank God, Jesus died on the Cross for *all* of our sins, including abortion. The Bible tells us in the story of David and Bathsheba that God has all our children who have passed on in Heaven, and there is forgiveness and reconciliation for all who are His!

If you've suffered abortion, no one—including Planned Parenthood—can honestly tell you that abortion is as "simple as going to the dentist." I know it is not: nine miscarriages later, suffering depression, self-medicating, and feelings of shame and guilt were just some of the aftermaths with which I dealt. . . . Statistics prove that most post-abortive women suffer these struggles—and more. I present these facts because I know firsthand the travesty of the abortion story. . . . Again, please consider all sides of your decision, because there is even more to know about the medical truth in regard to abortion: This is why I prompted my dear friend, Dr. Jack Monaco, to write his book *Second Touch*.

His professional account will lay out the spiritual truths, scientific facts, and medical findings that a baby in the womb, no matter the time of gestation, is, in fact, just that: *a baby*. From Dr. Jack's years of study to become a top-notch physician, followed by decades of delivering infants as an OB/GYN, he clearly tracks for you in his book the facts about the *life issue*. On a personal note, he also addresses God's Word on the matter: Life is sacred, and God knew each of us before He knit us together in our mother's womb (Psalm 139: 13–14).

As a Bible study partner of his, I quickly learned that Dr. Jack was a passionate supporter of the pro-¬life issue with firsthand knowledge that *everyone* considering abortion should know prior to making their decision.

Dr. Jack is a compassionate, learned man of God and a physician for *life*, so he has taken the time to write the truth about abortion. . . . *Please* read this book, and allow the wisdom found within its pages to become words written on your heart.

There are so many alternatives to abortion, starting with support by your local Pregnancy Centers which offer a sneak peek of your baby by way of an ultrasound, as well as emotional, financial, and practical support *beyond* the birth of your baby. And, of course, there is always *adoption*: Birth parents who sacrificially honor their baby's life and allow those waiting to adopt *are my heroes. Praise God,* what a gift!

Thanks, Dr. Jack, for writing *Second Touch*. I pray in agreement with you that the truth you share in your book will save many lives of the unborn.

Blessings,

"Jen" Jennifer O'Neill

Covergirl/Actress/Author/Speaker

# Preface

WHEN I GRADUATED FROM MEDICAL SCHOOL, I believed that as an obstetrician/gynecologist participating in women's health, it was my obligation to provide complete care for women. To that end, I considered myself to be pro-choice. I believed then that it was a woman's "choice" to carry or abort her pregnancy. In either case, it was my duty to help her to complete her decision safely. I took the safe route, absolving myself of any guilt over her decision by placing the entire burden on the patient.

The death of my teenage sister greatly influenced my choice in pursuing a medical career that focused on *life*, not death. Therefore, I became an OB/GYN, and it was the right choice for me.

One day, a patient came in for her annual exam and asked me if I was "pro-life" or "pro-abortion." I am not sure why we began such a conversation, but I realize now that this was God's way of reaching me through my patient. I told her that I was "pro-choice." She responded by saying, "So you are pro-abortion," which I strenuously denied.

But no matter how much I squirmed and spun my answer, I soon realized that she was right: Being "pro-choice" is being "pro-abortion." At that moment I became "pro-life" and have never looked back.

God had achieved his goal for my life through this woman. I hope that she may somehow read this book and know that I am eternally grateful to her for that epiphany.

I chose to train and practice in a Catholic hospital specifically so that I would not be forced to work in a "termination" clinic. I missed nothing by not doing terminations, and my training was more than complete in every aspect of women's healthcare.

I am eternally grateful to Jennifer O'Neill, who has become a dear friend and Bible Study partner, for asking me to write this book. Given my career as an obstetrician/gynecologist, the experience that goes with more than forty years of practice, and my Christian faith, writing this book was a natural and long overdue project.

The time has never been more critical than now to address the truth about abortion, especially in light of the US Supreme Court's recent reversal of *Roe v. Wade*. When you see the numbers of terminations done each year in this country, and look at the demographics regarding which segments of the population are seeking abortion, one begins to see the magnitude and impact abortion has on society. Historically, this is not a new story, but it's time for *truth* to be clearly told and for the community to respond to the reality of abortion and to recognize that it not only affects the baby, but also the entire family and society at large.

Most of the research for this book comes from the study of science, embryology in particular, and my own personal experiences attending to several thousand pregnancies and births. My hope is to help women and young ladies who are facing an un-expected pregnancy to realize that there are alter-natives available, and they are waiting for you and your baby!

— Dr. Jack Monaco

# Acknowledgments

F IRST AND FOREMOST, I AM GRATEFUL to my Lord and Creator who has given me the gifts and opportunity to achieve all that I have as a physician, obstetrician, husband, and father.

To my wife, Lesa, for her undying support, encouragement, friendship, and love which has helped me complete this journey.

To my patients, who have given me the opportunity to assist them in bringing into the world the new life that has been created within them.

Lastly, to my dear friend, partner in Christ, and my editor Jennifer O'Neill, whose encouragement, advice, and support has made all this possible. Without her prompting me to write about my experiences as an obstetrician, this book would not have been written.

# Introduction

WHEN DOES LIFE BEGIN? This question has been and will continue to be a matter of discussion. I wrote this book to tell you the *truth* about abortion from both a medical as well as a spiritual point of view. My hope is that you, the reader, will embrace the idea that everyone has the right to life. As a Christian and physician, I believe that life does, indeed, begin at conception. . . . My heart tells me that. So do medical and scientific facts.

However, if you're not sure, I encourage you to continue reading and take an honest look at both sides of the issue. I ask you to open your heart and mind to receive the facts, and to consider the eternal impact of the "Pro-Life" vs. "Pro-Choice" debate.

The fact is, everyone was conceived and nurtured in their mother's womb until their birth—when they received their God-given inalienable right to life—as guaranteed by our Constitution.

Since you are reading this book, you are obviously enjoying that right. But what about the sixty-million-plus aborted babies (just in the United States alone) who had that right to life taken from them? While we cannot bring those babies back, we do have the moral obligation, based on irrefutable information and available opportunities to protect unborn children in the future.

As an obstetrician, I have attended thousands of women in labor while experiencing the miracle of birth during the delivery of their babies, so I speak from personal and medical experience. God's First Touch is the creation of life, and my being the "second" one to touch that life makes each delivery special. Helping to bring a new baby into the world, and holding that newborn child, is the greatest thrill that I have ever experienced.

My deep desire for you is that you will be able to embrace, through my experiences, the sacred wonder and sanctity of life.

# SECOND TOUCH

# OBSTETRICS: MY STORY

# Chapter 1
# Obstetrics: My Story

WHEN I WAS IN SEVENTH GRADE, my older sister, Barbara, died unexpectedly at the age of eighteen from leukemia. She had struggled with ulcerative colitis, an inflammatory condition of the large intestine, since age twelve. Obviously, her death had a significant adverse effect on me as a child that persists to this day. I began to think about mortality and dying at an age when all kids my age wanted to do was play baseball, ride bikes, and have fun. It wasn't until I came to my faith in Jesus Christ that I became comfortable with the notion of dying someday.

Barbara was diagnosed with leukemia just two months after graduating from high school. Treatment for childhood leukemia was in its infancy. Compared to today, there was not much that could be done. She was placed on the waiting list at St. Jude's Children's Research Hospital—a waiting list that proved to be too long for her. She died two months later.

Losing Barbara had a devastating effect on me. I never got the chance to say goodbye to my sister before she died, and my parents would not let me see her after her death. I was "shielded" from the pain of a wake and funeral, and I was never able to get closure: My parents divorced a few years later.

After Barbara's death, I became fascinated with science. At the age of twelve, I felt called to study medicine. My driving focus was to find a cure for cancer, specifically leukemia. Realizing that Barbara's death caused me to dwell on death and my own mortality, I decided to pursue a career in medicine, specifically the specialty of life—obstetrics and gynecology.

My faith in Jesus Christ helped me to focus on a new mission of hope and not one of helplessness.

I began volunteering at a local hospital at the age of fourteen. Once old enough to work, I was hired in the admitting office doing clerical work. Before long, I was able to aid patients as a nursing assistant, and I eventually worked as a surgical tech in the operating room.

Despite my resolve, my pursuit of a career in medicine was a hard-fought challenge. As a college freshman, one of my first requirements was dissecting a fetal pig in biology. For those who have never seen a fetal pig, it is a fully formed pig just days away from being born, when the mother pig is slaughtered and destined for the grocery store.

My pig looked like a regular pig, only smaller, having fully formed ears, hooves, and snout. It even had eyelashes. I could almost hear it squealing as it lay in the dissecting tray before me. I sat at my seat with this dead pig in front of me for what seemed to be an eternity before I could muster the courage to make the initial incision to begin the dissection that would expose the abdominal contents. Truth be known, I think that I was the last one in the class to get started, all the while knowing that if I could not make that incision, I would never be able to have a career in medicine.

When I finally was able to make that incision and observe the abdominal contents, it was then that I realized that no human or computer could so perfectly design such a marvel of creation. Only God could master a plan of such intricate design that meshed form and function into a complete living being; in this case a pig. I found it ironic that my first experience with the miracle of life was by studying something after death.

Although I have always been a good student, I had to work especially hard to gain acceptance into medical school. You see, the war in Viet Nam was in full swing and applications to medical schools across the country were at record highs. Fortunately, an extra helping of determination (likely resulting from growing up in New Jersey, where the words "no" and "can't" weren't a part of my vocabulary), gave me the drive to succeed. It wasn't until years later that I realized that God played a huge part in my being in medical school by putting people into my life who helped me to fulfill my dreams.

One such person was a family friend who owned a surgical instrument manufacturing company. He kindly gave me several instruments including scissors, needle holders, and forceps. By this time, I was working in the operating room of a local community hospital, so I could use the instruments to practice sewing and tying knots at home. The surgeons, with whom I worked, also took an interest in teaching me how to correctly hold and use these surgical instruments, how to tie knots, and even let me do a bit of sewing. So, with a lot of practice, by the time I was in my clinical rotations, I could assist and sew at least as fast as the residents in surgery.

I graduated from Wake Forest University (Bowman Gray) School of Medicine. It was an exceptional four years (actually five for me—I did a year of cardiovascular research in the physiology department). Our class was unique. Unlike some stories told about competitive, self-centered students at other medical schools, we helped each other to succeed, not fail. We realized that our goal was to be the best doctors that we could be; helping each other to achieve that goal was important. It was evident that the grace of God was present in both the faculty and philosophy of the school, and it continues to be to this day.

Obstetrics was a natural choice for me. After all, besides carrying a life in the womb, what could be more special than delivering babies and bringing life into the world?

The next thirty years were full of blessings, difficulties, wonders, stresses, hard work, and many sacrifices. In retrospect, I had an awesome career as an OB/GYN, and I would not have done anything differently.

Those years have given me a perspective on life that I don't think I would have in any other profession. I realized that life is not only a precious gift but is truly a miracle in motion. A single cell contains all the genetic signals to produce an entire living human being comprised of trillions of specialized cells, and this happens in only nine months (40 weeks, 280 days!) I find it difficult to believe that, given the facts of life as evidenced by the joining of the mother's egg and father's sperm, that anyone can deny the truth that life begins at conception.

# FROM THE BEGINNING

# Chapter 2
# From the Beginning

MEDICAL SCHOOL STUDENTS were treated with respect by the residents and faculty during my time there. Our teachers showed a genuine interest in teaching and helping us develop into good and caring physicians. They did this by stressing patient care and the "human side" of being a doctor. We were exposed to patients very early on to help us establish a "bedside manner" that is sometimes difficult for some doctors to develop. As my former partner said, "There is the science of medicine, learned through years of study, but it is the art of medicine that is learned through a lifetime of practice."

At Wake, we got a head start on the art of medicine.

During my early years of training, I was always astounded by the incredible joy surrounding the birth of a child. But it wasn't until years later that it became crystal clear to me that life truly begins at conception.

As I mentioned previously, I purposely chose to train at a Catholic hospital specifically to avoid having to do abortions. Unfortunately, one night, I was placed in a situation personally and morally, a challenge to me as the chief resident on duty. I received a call from one of the attending physicians notifying me that he was sending a patient to the emergency room. He was a local "abortion doctor" who had started a termination procedure on a patient whom he believed was in her first trimester (less than twelve weeks), but she was in fact closer to twenty weeks pregnant. He sent the patient to the hospital for the resident staff (me) to complete the abortion. Since I was the chief resident and, at that time, the resident staff could not refuse an order by an attending physician, I had no choice but to take the patient to the operating room and complete this abortion. At this point it was about taking care of the patient.

To say that it was a horrible experience does not begin to describe the surgical experience that followed. I told the nurse, who was to assist me, that it would not be pleasant and that she did not have to stay. To her credit, she stayed to help. A nearly twenty-week fetus is quite recognizable as a baby. (After all, the fetal heart begins to beat by twenty-one days of development.) Fortunately, times have changed, and the resident staff is no longer placed in situations such as this.

On another occasion, I was on my way to a radio station in Hartford, Connecticut, where I was scheduled to be interviewed by a well-known and very popular radio personality. Brad Davis was an ex-Marine and a believer. On my way to the station, I was listening to the guest on his show who he was interviewing before my time slot. This guest was talking about cloning. Although Brad was impressed, he was also a bit incredulous. He commented to his guest, "Are you saying that you can make an entire Brad Davis from one of my cells? His guest affirmed this.

During the break after my first segment on his show, he asked me what I thought about this cloning issue. I replied that if one believed that you could clone an entire organism from a single cell, you cannot deny that life begins at conception. I will never forget the look on his face as I explained that the union of a single sperm and egg contains all the genetic information (a complete blueprint of life) to begin the complicated process of turning a single cell into trillions of specialized cells that make a living human being. This is no accident! The "Miracle of Life" starts with a single fertilized cell.

# THE DEVELOPMENT
# OF THE FETUS

# Chapter 3

# The Development of the Fetus

As God's timing and direction would have it, my very first clinical rotation in medical school was OB/GYN. I had hoped to have more clinical experience before starting with my chosen profession. It was a dramatic introduction to the world of surgery; delivering babies, long hours of hard work, sleepless nights, and the pressures of being a "doctor" heralded to save lives! We worked thirty-six hours and had twelve hours off. The medical students were on call more frequently than the residents who took calls every third night. We were "on" every other night.

After my third night out of five nights without sleep, I was so exhausted, I actually considered leaving medical school and finding another profession, but I decided to give it a bit more time. I'm glad that I did.

It was about this time that ultrasound technology came into the world of clinical medicine. It was fascinating to actually "see" into the uterus and watch a living fetus move inside the womb of the mother. Imagine, a "life within a life!"

After "seeing" life, I was reminded of my course in embryology in which we studied the beginnings of development from the fertilized egg through more advanced stages of development. I spent a lot of time staring into a microscope and looking at the changes that are orchestrated by the genetic messages that begin with fertilization. It was amazing to see the beating heart of a chick embryo after only a couple of days of development.

It became apparent very quickly that to take a single cell to completion was an amazingly complex and well-orchestrated process of development. In fact, it is miraculous that this process happens every minute of every day (in most cases) without errors. And while errors do occur, they are relatively rare and usually have an explanation. I look forward to asking God about such things when I meet Him in person.

Until then, however, I ask you to consider the following: Even the process of a single sperm reaching an egg is a complex and incredible event. Imagine a single cell that has the ability to "swim" through the uterus into the fallopian tube to the waiting egg. This sperm has no heart or lungs, no brain or feelings, yet it has the ability to generate the energy needed to power a "tail" which provides movement to follow a chemical trail that leads it to the egg. Sperm also contains one half of the genetic information from the father, which, when combined with the other half of the genetic information from the mother, initiates the process of life. These facts will never cease to amaze me!

Just to repeat, because this is very important and amazing, the human embryo's heart begins to beat at twenty-one days of development. By eight weeks, the head constitutes almost half of the baby's body. By twelve weeks, the body has almost doubled in size, and the growth of the head slows. The eyes and ears are recognizable, and the arms, legs, hands, and feet have developed. This period of development is often called the stage of initial activity because the baby reacts to stimuli. . . . That is pain and touch![1] These movements are, however, too slight to be felt by the mother. Quickening, or the first feelings of movement, doesn't occur until the seventeenth to twentieth week in most cases. By the end of the twelfth week, sucking movements by the baby can occur.

Further growth and development continue, and by twenty-six to twenty-nine weeks (six and a half to seven months) the child can survive even if born prematurely. The ability to survive a premature birth is due primarily to the development of the fetal central nervous system (and the scientific advances in neonatal care). At this stage, the fetal nervous system is able to direct rhythmic breathing movements and control body temperature. The eyes open during this period. Growth begins to slow three to five weeks before the anticipated date of birth.

The complexity of the baby's development from fertilization to delivery is clearly a miracle. It is difficult to fully comprehend the amazing process that God has orchestrated for us-and for all life He created. There is no human or computer program that could have designed a process such as this. To think otherwise is just plain nonsense. We have a powerful and loving God who has created all life, which, if protected, is able to thrive for His glory, praise, and service!

# PERSONAL ACCOUNTS

# Chapter 4
# Personal Accounts

A S ANY WOMAN WILL ATTEST, the relationship be-tween an obstetrician and his/her patient is both intimate and special. There is a level of trust and confidence that seems to exceed other specialties. That is not to say that other physicians don't have unique relationships with their patients, but rather that the patient/gynecologist bond is indeed an exceptional experience.

I have always believed that sharing personal stories and events with my patients gave a more caring aspect to my practice. I especially liked to inject humor, where appropriate, in various situations. But all that said, I have been beyond honored to be the

"Second Touch" to a mother's most treasured moments.

On a personal note: Most obstetricians vividly remember their first delivery. We are cautioned about how slippery a newborn baby is, and we practice how to not drop the child. Most obstetricians during their early training have had "near misses."

While I consider all deliveries to be special, some are especially memorable: After the infant is delivered, identification bracelets are placed on three extremities, and the baby's footprints are taken. I had just delivered a breech baby, the third infant of a young couple. Both parents were grateful for a successful outcome (and so was I). They also both had a good sense of humor. The mother happened to comment on the footprint process. I suggested that we also get "prints" of the part of her daughter that presented first. Her husband assisted me as we took "butt" prints on a separate "unofficial" birth card. Then I suggested to Mom & Dad that they make reduced-size copies of the "butt" prints and use them as birth announcements.

While most of my deliveries were something to celebrate, there are those rare and tragic deliveries of developmental anomalies that I attended. Probably the most heartbreaking occurrence for an expectant mom and her family is the death of the baby at term, especially if it is the woman's first pregnancy. It is incredibly difficult to have to go through an entire labor and delivery knowing that your child has already died in the womb. The sorrow and grief is enormous. Fortunately, I have had to attend relatively few of these deliveries. There are no words of consolation that can lessen the pain those parents feel.

Clearly, the loss of a child to abortion is no less devastating.

As my career advanced, I found myself attending women in labor whom I had delivered years before. It is the ultimate compliment to an obstetrician to care for "families" and to be involved in multiple generations of moms.

Over the years, I have received many words of appreciation from grateful patients, but one gift in particular stands out. I had four patients who happened to know each other as well as work together. They all delivered their babies within several months of each other. One day, all the moms came to my office with a collage of pictures of them holding their children, all of whom I had delivered. I can't think of a greater honor than having moms express their gratitude in such a thoughtful way.

We are all gifts of our Creator, and we are here to serve a purpose, often greater than we realize.

# THE MIRACLE
# OF LIFE

# Chapter 5
# The Miracle of Life

TO BELIEVE IN A LOVING, CARING, ALL-POWERFUL GOD is not something everyone claims. There are many examples of people who were not believers researching the evidence for Christ. Personally, I find it difficult to ignore the many examples of God's presence in our lives.

I have many friends who do not share the same belief we Christians profess. I've noticed that, in some way, each of them seems to be "missing something," or struggle in their lives. While being a Christian certainly does not guarantee a "perfect life," it does give us a different perspective of how to live our lives, and how to accept the difficulties and challenges that we will all experience.

I was raised to respect people, to live my life in such a way that I would be able to benefit others, and most of all, to respect life. We all know that currently there is an overwhelming increase in personal attacks both physical and verbal, a confusion of "identity," and a general lack of respect for others. I recall a news story of a young child who was brutally murdered for his new sneakers—and then there is the most recent war on Israel by terrorists.

Frankly, none of this surprises me. If we have so little value for the unborn life, why would it be a shock that there is also such little regard for any other life, no matter the age of the victim?

A popular TV show about a family of police officers in New York City talks about the "broken window" theory. A small event, like a broken window, if overlooked and not punished will eventually lead to bigger crimes against society. If we overlook the killing of the unborn child, it is only a matter of time before we become indifferent to other more serious crimes and begin to "give the 'perp' a break" because of his or her childhood trauma or less than ideal upbringing. No matter how you spin it, abortion is murder, a crime against humanity, and an insult to the Creator of Life!

But, as I have said before, the aborted baby is not the only victim.

Not long ago, I found myself in a discussion with a woman who told me that she was a "liberal." With that declaration, I asked her for clarification of something that I was trying to understand; namely, how one could favor abortion, but not be in favor of capital punishment for a violent criminal who had killed another living human being? The look that she gave me is one that I will never forget. She answered, "If you're referring to abortion, I know that if I were raped, no one should have the right to tell me that I can't get rid of that 'thing' growing inside of me!" I asked her if she knew that the "thing" of which she spoke is a human life, regardless of how the pregnancy occurred. I asked her if she knew that only 1-2 percent of females who are raped choose abortion. That statistic also holds true for female victims of incest who become pregnant.

I went on to say to her that if that is the case, perhaps we should amend *Roe v. Wade* to apply only to cases of rape and not to all unintended or unwanted pregnancies. I challenged further by asking her if she was for abortion as a form of birth control. She looked at me in shock. I explained that for all those mothers of rape or incest babies who chose LIFE, the reason for their resolve in those cases was because they did not choose to perpetrate the same violence on the baby in their womb that was forced on them as victims. The discussion ended by my sharing to her that whether the moms decided to raise their baby or to find a loving family through adoption, the innocent life was saved.

As a physician, we are expected to preserve and maintain life, to relieve pain and suffering, and to improve our patients' quality of life. Somewhere along the way, we have strayed from our Hippocratic Oath, which says, among other things: "*Primer non nocere* —First, do no harm." If you believe that abortion is "doing no harm," then nothing I say will change your view. However, if you look at the larger picture, by supporting abortion, society is overlooking the "broken window" and accepting this seemingly "simple procedure" as a minor infraction devoid of far-reaching consequences. I can't help but wonder how many professors, scholars, Nobel laureates, future presidents, doctors, or lawyers have been aborted! Only the Lord knows.

# THOUGHTS
# TO PONDER

# Chapter 6
# Thoughts to Ponder

THERE IS ONE THING THAT ALL OBSTETRICIANS HAVE: a lot of time waiting for their patients to deliver. That is time to think, and time to ponder. Yes, I spent many countless hours waiting and thinking about the basic desires of life: Will I make it to my son's hockey game or my daughter's ice-skating show? Have I saved enough for their college education? Who will they marry, and will they be happy?

I was always thankful for the gifts that I had been given that allowed me to become a successful doctor, surgeon, and father. No matter how routine or mundane, each delivery I attended, each life that I helped to bring into the world, was a gift to the parents and a gift to me. Being an obstetrician is a special calling—a most unique profession.

It always puzzled me that a doctor of "life" could also perform a procedure ending life (abortion). While it can be argued by some that it is part of providing "complete women's care," it is still the antithesis of life, life that we swore to protect. Some might believe this is a naive "Pollyanna" mentality. No matter, it is still life vs. death.

Life is short and precious. It is a gift on multiple levels. Our natural instinct is to fight for life. I firmly believe that life starts at conception and that we should not be so quick to destroy another human being's future for any reason.

God does not make "mistakes" in His process and purpose of His sovereign creations. There are numbers of crisis pregnancy centers all across America just waiting to help you or someone you love *choose life* for their unexpected *baby*. They are waiting to support that mom and family in *every* way they

have need. Please reach out to them. They are waiting for you with unconditional love, options, support, prayers, and follow-through for you and your baby.
Blessings, Dr. Jack

# APPENDICES

# Appendix A
# Abortion Laws by State

IN JUNE 2022, THE U.S. SUPREME COURT overturned *Roe v. Wade*, thus opening the door for states to ban abortion outright. As of early 2025, abortion is illegal in twelve states. The state-by-state breakdown regarding abortion's legal status follows. Each list is arranged alphabetically.

*(Source: Center for Reproductive Rights, https://secure.reproductiverights.org/):*

**BANNED (Enforcement ban criminalizes abortion):**
Alabama, Arkansas, Idaho, Indiana, Kentucky, Louisiana, Mississippi, Oklahoma, South Dakota, Tennessee, Texas, West Virginia.

**HOSTILE (Enforces bans on abortions after 6-12 weeks [state specific]):**
Florida, Georgia, Iowa, Nebraska, North Carolina, North Dakota, South Carolina, Utah, Wisconsin, Wyoming.

**NOT PROTECTED (Accessible without legal protection):**
New Hampshire, New Mexico, Puerto Rico, Virginia.

**PROTECTED (Allowed with legal protection):**
Alaska, Arizona, Colorado, Delaware, Kansas, Maine, Massachusetts, Michigan, Missouri, Montana, Nevada, Ohio, Rhode Island.

**EXPANDED ACCESS (Legal with expanded access to services):**
California, Connecticut, Hawaii, Illinois, Maryland, Minnesota, New Jersey, New York, Oregon, Vermont, Washington

# Appendix B

# Pro-Life Counseling Resources

ON THE FOLLOWING pages is a list in no particular order of twelve facilities, organizations, or associations that offer assistance and/or counseling to women who have an unwanted pregnancy and/or have had an abortion and seek the types of services shown below. These twelve entities are ones that I have either been personally involved with or have been referred to me. You are welcome to search for more if what you believe would best suit you is not listed here. The information presented here was taken from their respective websites.

## 1. ABC Women's Center
https://abcwomenscenter.org

Two locations:
8180 East Main St., Middletown, CT 06457
860-344-9292

8 Concord Street, New Britain, CT 06053
860-344-9292

Mission Statement: "An unexpected pregnancy can be one of the many reasons why life gets complicated. Perhaps this is not the right time for you to start a family. Or, maybe it's the wrong guy, or you are struggling to make ends meet financially. Regardless of where you are right now, you still have a decision to make. Whatever you choose won't be easy, but that is why we are here to help you through this process."

Services offered:
- Parenting program (Baby items, formula, baby food, etc.).
- Post-abortion support.
- STI/STD information.
- Community referrals for services not offered.
- Social media blog with FAQ answers.

## 2. Prolife Across America

https://prolifeacrossamerica.org
800-366-7773
*"Se Habla Espanol"*
P.O. Box 18669, Minneapolis, MN 55418

Services:

- Pregnancy resources.
- Abortion pill reversal.
- Adoption resources.
- Post-abortion resources.

## 3. The National Right to Life Foundation, Inc.

https://nrlc.org

Founded in 1968, National Right to Life is the nation's oldest and largest pro-life organization. National Right to Life is the federation of 50 state right-to-life affiliates, the District of Columbia and more than 3,000 local chapters. Through education and legislation, National Right to Life works to restore legal protection to the most defenseless members of our society who are threatened by abortion, infanticide, assisted suicide, and euthanasia.

Mission Statement: "The mission of National Right to Life is to protect and defend the most fundamental right of humankind, the right to life of every innocent human being from the beginning of life to natural death."

Key points:
- Oldest and largest pro-life organization in the United States.
- Mission to protect and defend the rights of the unborn.
- Dedicated to advocating for the sanctity of human life from conception to natural death.
- Works to promote legislation to safeguard the rights of the unborn.

## 4. The Susan B. Anthony Pro-Life America
https://sbaprolife.org
1-800-7124357

- Focuses on electing pro-life candidates to public office.
- Seeks to combine women's rights and the right to life.
- Works to provide resources, training and support to candidates and lawmakers committed to defending the sanctity of life.

## 5. The March for Life
https://Marchforlife.org

The March for Life is an annual event, but pro-life education and advocacy is important all year long.

Key points:
- Organizes an annual "March for Life" in Washington, DC.
- Committed to promoting the dignity and protection of every human life, especially the unborn.
- Peacefully demonstrate their support for life and urge lawmakers to enact pro-life legislation.

## 6. Students for Life of America
President: Kristan Hawkins
Executive VP: Tina Whittington (twhittington@students for life.org)
https://studentsforlife.org
1000 Winchester Street, Suite 301
Fredericksburg, VA 22401
540-834-4600

Description: Pro-life organization dedicated to empowering and mobilizing young people in the pro-life movement.

Mission: To educate and activate the next generation of pro-life leaders.

## 7. Human Life International (HLI)
https://www.hli.org
4 Family Life Lane Front Royal, VA 22630
800-549-5433
Info: li@hli.org

Offers online informational services addressing:
- Abortion
- Birth Control Population Control
- Marriage, Family & LGBTQ Issues
- Faith and Catholic Family Life
- End-of-Life issues
- Assisted Reproduction
- Podcasts
- News
- Free E-books

Services:
- Leadership training: Priests, doctors, teachers, and other leaders have a huge influence at the local level. HLI missionaries train individuals from all these groups.

- Chastity Education: Contraception and graphic sexed programs create a culture of casual sex. When contraception inevitably fails, people look for abortion.
- Pregnancy Care Centers: Thousands of children lose their lives to abortion every day. Mothers all over the world feel like it's their only option. These women need to hear that there are alternatives to ending the life of their child.

Key points:
- Global pro-life organization committed to promoting and defending the sanctity of human life worldwide.
- Provides educational programs, resources and initiatives to advance a culture of life.
- Promotes the principles of Catholic teaching on life issues and actively opposes.

## 8. Americans United for Life (AUL)
https://aul.org
1150 Connecticut Avenue NW, Suite 500
Washington, DC 20036
202-289-1478
Info: Life@aul.org

Mission: We advance the human right to life in culture, law and policy.

Our Vision: We strive for the day when all are welcomed throughout life and protected in law.

Our Method: We equip advocates and lawmakers with the facts and strategies that change hearts and minds and protect human life.

Our Motivation: We believe all human rights flow from the human right to life. We are all equal members of the human family and equally worthy of respect, solidarity, and love.

Key points:
- Dedicated to protecting the sanctity of human life thorough advocacy, litigation, and policy work.

- Provide legal expertise, resources, and guidance to lawmakers and activists.

- Protect life at all stages, advance pro-life legislation and foster a culture that respects the inherent dignity and value of every human being.

## 9. Care Net Pregnancy Center

https://care-net.org
44180 Riverside Parkway, Suite 200
Lansdowne, VA 20176
703-554-8734

Mission: Care Net centers understand that pregnancy decisions are an emotional and private choice. Clients can feel confident in the information they receive because it's backed by research from caring professionals. Affiliated pregnancy centers do not discriminate based on age, race, nationality, religious affiliation, disability or any arbitrary circumstances. Care Net affiliated pregnancy centers do not perform or refer for abortion and do not profit from any client's decision.

Key points:
- Network of pregnancy resource centers across the United States.
- Provide compassionate support and resources for women facing unplanned pregnancies.
- Offers free pregnancy tests, ultrasound, counseling, and material assistance.
- Empowers women to make informed decisions about their pregnancies, providing nonjudgmental support.

## 10. Heartbeat International
President: Jor-El Godsey
Board Chairman: Peggy Hartshorn, PhD
https://heartbeatinternational.org

8405 Pulsar Place, Suite 100
Columbus, OH 43240
Phone: 614-885-7577
Info: Support@heartbeatinternational.org
Mission: Advancing life-affirming pregnancy help world-wide

Key points:
• Global network of pro-life pregnancy help organizations and medical clinics with over 3,600 pregnancy help locations in more than 90 countries

• Provides support, resources, and training to pregnancy centers and medical professionals worldwide.

• Focuses on promoting alternatives to abortion, empowering women to make informed choices.

## 11. The National Institute of Family & Life Advocates (NIFLA)
https://nifla.org
10333 Southpoint Landing Blvd.
Fredericksburg, VA 22407
540-372-3390
Info: admin@nifla.org

Mission: The National Institute of Family and Life Advocates (NIFLA) exists to protect life-affirming pregnancy centers targeted by pro-abortion groups and legislation. Through legal counsel, education, and training, NIFLA enables member centers to avoid legal pitfalls in their operations.

Medical: NIFLA recognized the importance of using ultrasound in a pregnancy center setting for reaching abortion-minded women. Ultrasound offers a window to the womb, and this impacts a woman's decision to choose life.

Key points:
- Dedicated to supporting and providing legal assistance to pregnancy centers and pro-life medical professionals.
- Works to protect the rights of these organizations and professionals to operate in accordance with their pro-life beliefs.
- Offers training, resources, and legal expertise to help pregnancy centers navigate the legal challenges, and provides effective care to women facing unplanned pregnancies.
- Strives to ensure that pregnancy centers can continue to offer life-affirming options and support to women and families.

## 2. Save One

Founder/President: Sheila Harper
https://saveone.org
114 Canfield Place, Suite B-6
Hendersonville, TN 37075
615-636-2654
Info: info@saveone.org

Mission: Abortion has a deep ripple effect. You may have chosen abortion personally, lost a child to abortion, or your life has been profoundly affected by abortion. SaveOne can help you through the healing process.

# Appendix C
# Statistics

W HAT FOLLOWS ARE MANY STATISTICS AND COM-MENTARIES that will illustrate the magnitude of the effect that abortion has on not only the woman, but also society as a whole. These effects are personal, emotional and financial in their scope.

The Pew Research Center conducted many surveys about abortion over the years. In a survey conducted nearly a year after the Supreme Court's June 2022 decision that overturned *Roe v. Wade*, 62 percent of U.S. adults surveyed said that abortions should be legal in all or most cases, while 36 percent said that it should be illegal in all or most cases.

The PEW Research Center compiles surveys and statistics showing changes and trends from year to year.

An exact number is hard to come by. The CDC (Centers for Disease Control and Prevention) and Guttmacher Institute have both tried to measure this for nearly fifty years using different methods and publishing different figures. Following is what I have been able to compile from those sources:

For 2021, The CDC reported a total of 625,978 abortions for the District of Columbia and the forty-six states that has available data, up from 586,355 in 2020. The corresponding figure for 2019 was 607,720. (Many states do not report abortion statistics.)

In 2020, the Guttmacher Institute reported a national yearly total of 930,160 abortions covering all fifty states and the District of Columbia.

The annual number of U.S. abortions rose for years after *Roe v. Wade* legalized abortion in 1973, reaching the highest levels in the late 1980's and early 1990's, according to both the CDC and the Guttmacher Institute. Since then abortions seem to be decreasing at a "slow yet steady pace."

In 2020, there were 14.4 abortions per 1,000 women ages 14-44 (29.3 per 1,000 in 1981). This number decreased in 2021 to 11.6 per 1,000.[2] (This excludes data from California, DC, Maryland, New Hampshire, and New Jersey.)

The national rape-related pregnancy rate is 5% per rape among victims of reproductive age (12-45). Only 11.7% received immediate medical attention after the assault; 47.1% received no medical attention at all. 32.4% of these victims did not discover that they were pregnant until they were in the second trimester. Of these, 32.2% opted to keep the infant, whereas 50% underwent a second-trimester abortion and 5.9% placed the infant for adoption. 11.8% had a miscarriage.[3]

It is currently estimated that there are 1,603 abortion facilities in the U.S. This includes 807 clinics, 530 hospitals, and 266 physicians' offices. The majority of abortions are medical (pill induced).[4]

The majority of women (57%) having abortions were in their 20's while (31%) were in their 30's. Teens ages 13 to 19 accounted for 8% while women aged 40-

44 accounted for about 4%. The vast majority of women having abortions were unmarried (87%), with 13% being married.[5]

According to CDC data, 42% of women having abortions were non-Hispanic Blacks, while 30% were non-Hispanic white, 22% were Hispanic, and 6% were other races.

For 57% of the women having an abortion in 2021 it was their first time; 24% were second time; and 11% were third-time abortions. For 8% it was their fourth time or more. The vast majority (93%) of abortions occur during the first trimester of pregnancy (less than 13 weeks), while 6% occurred between 14 and 20 weeks of gestation and about 1% were 21 weeks or more.[6]

Roughly 2% of all abortions involve some complication for the woman. Most are minor (pain, bleeding, infection, and post-anesthesia complications). According to CDC "case-fatality rates," there were 0.45 deaths per 100,000 legal induced abortions. In 2020, the CDC reported that 6 women died due to complications of an induced abortion, compared to 4 in

2019, 2 in 2018 and 3 in 2017. Since 1990, the annual number of deaths among women due to a legal abortion has ranged from 2-12. The number is higher when counting both legal and illegal abortions.[7]

**Author's Commentary:** There are voluminous statistics looking at nearly every aspect of abortion from the demographics (which socioeconomic groups have the highest rate of abortion to maternal complications of abortion).

## Pregnancy and Rape/Sexual Assault

One in 20 million women in the United States have reported an experience of a pregnancy from rape, sexual coercion, or both in their lifetimes. More than 3 million women have experienced pregnancy resulting from sexual coercion in their lifetime; of these, 28% experienced a sexually transmitted disease and 66% were injured. More than 80% of those who were coerced were fearful or concerned for their safety. Non-Hispanic multiracial women experienced higher rates of rape or sexual coercion compared to all other groups. Women raped by a current or former intimate partner were more likely to report rape-related pregnancy (26%) than those raped by a stranger (6.9%) or an acquaintance (5.2%). These findings illustrate the connection between sexual violence, intimate partner violence, and women's reproductive health.[8]

Ninety-five percent of abortions are the result of unintended pregnancies. The most common reason for unintended pregnancies is failure to use contraception (52%) or inconsistent/incorrect use of contraception (43%). Common reasons for abortion include economic, social, emotional, and family issues. Medical reasons include life-threatening maternal health conditions and fetal malformations that are inconsistent with life outside the womb.

**Author's Commentary:** The debate continues regarding a women's options (right to choose) in the case of sexual assault. Very little attention is directed to offering the infant up for adoption. It may be seen as a punishment to the mother to make her carry a child, conceive by assault, to term and to go through the pain of delivery.

# APPENDIX D

# Planned Parenthood: Facts & Fiction

P LANNED PARENTHOOD FOUNDER MARGARET SANGER was born into poverty in 1879 to a large family, being one of eleven children. Her mother died from tuberculosis when she was just a young woman. Sanger blamed her mother's death on the fact that she had eleven children and the fact that they were impoverished. She felt that she could diminish poverty by preventing certain groups of people from having babies. This is the underlying theme of Planned Parenthood, as noted by writer Susan Ciancio, in a 2023 article she wrote for *Human Life International.*[9]

"Sanger eventually went to nursing school but

never finished. Instead, she began to work with impoverished women in New York City," Ciancio writes. "About this time, she joined the Women's Committee of the New York Socialist Party and began to advocate—in both her speech and writing—for sex education and birth control, even starting her own journal to reach as many people as she could with her ideas.

"Sanger explained her reasonings for why she believed that women 'need' birth control. In *Birth Control and Racial Betterment*, Sanger writes: 'We who advocate birth control … lay all our emphasis upon stopping not only the reproduction of the unfit but upon stopping all reproduction when there is not economic means of providing proper care for those who are born in health.' "Her articles and speeches are riddled with beliefs such as this. … (In) 1921, she established the American Birth Control League which eventually took the name Planned Parenthood."[10]

The very name, "Planned Parenthood," gives the impression that its mission is to help couples "plan" their families, which leads the reader to assume that this is accomplished with contraceptive and pregnancy counseling. In fact, while these are listed as provided services, Planned Parenthood's major service is pregnancy termination. Planned Parenthood's annual report for 2023-2024 lists the following for that fiscal year:

2.08 million patients

9.4 million services

364,600 Pap test & breast exams

2.2 million birth control services

5.1 million STD tests & treatments

402,200 abortions[11]

While these numbers seems to indicate that Planned Parenthood provides a variety of "healthcare" services for women, a closer look reveals a different picture. You must understand that Planned Parenthood counts each service that they provide during a single visit as a separate service. So a PAP smear, contraceptive counseling, STD counseling, and STD testing each will be counted as "separate" services even though they were performed for the same person during a single visit. This skews their numbers—in effect it "dilutes" them—such that the numbers of abortions performed by PP show up as a lower-than-actual percentage.

Planned Parenthood's 2023-24 Annual Report also shows total assets reported for Planned Parenthood Global & Affiliates as exceeding $3.1 billion. Taxpayer funding in the form of Government Health Services Reimbursements, contracts, Medicaid Reimbursements and Grants is reported as accounting for 39 percent of their income.[12]

By law, taxpayer money cannot be used to fund abortions. Since 1977 the Hyde Amendment has banned such use of federal funds, the only exceptions being for termination of pregnancies that endanger the life of the pregnant person or that result from rape or incest.[13] However, a significant amount of their revenue comes from government sources, and Planned Parenthood can use it however they desire. According to Jeanneane Maxon, writing for the Lozier Institute, "In 2022-23, Planned Parenthood received an estimated $233,293,400 in revenue from various sources for abortions performed." While Planned Parenthood is considered a "nonprofit" organization, with annual revenues exceeding $2 billion, it prioritizes abortion. "Any form of government money to Planned Parenthood only serves to support its advocacy and provision of abortion. In just four years, Planned Parenthood's overall reported revenue has increased by over $384 million."[14]

Planned Parenthood does offer other services including men's and women's health services such as pregnancy testing, general healthcare, STD testing and treatment and vaccines, emergency contraception ("morning after pill", "Plan B", "abortion pill"). It also offers birth control but fails to mention the potential harm associated with birth control pills.[15]

As an Obstetrician/Gynecologist, I have seen many adverse effects of oral contraceptives including high blood pressure, deep vein thrombosis (DVT), cardiovascular events including stroke, and an increase in breast cancer as documented in the literature. I stopped routinely recommending oral contraceptives more than twenty years ago. I also stopped recommending the medicated IUD for the same reasons. Unfortunately, this leaves women with few viable choices: the non-medicated IUD (copper IUD), barrier methods (diaphragm and condoms), abstinence or permanent sterilization which is not a likely choice in this population. More unfortunately, abortion is all too often used as a method of birth control.

So, what happens if a woman is raped? According to the *American Journal of Obstetrics & Gynecology*, "the national rape-related pregnancy rate is 5% per rape among victims of reproductive age (12-25). . . . 32.4% of these victims did not discover that they were pregnant until they were in the 2nd trimester. Of these 32.2% opted to keep the infant, whereas 50% underwent a 2nd trimester abortion, 5.9% placed the infant for adoption and 11.8% had a miscarriage."[16] While rape is a serious and significant issue, it represents a relatively small number of total unplanned pregnancies and more

Women want to have the choice to abort or maintain the pregnancy. My body . . . My choice. You could argue that you have an additional choice not to abort and offer the child for adoption. As a Christian, being Pro-Choice, in the traditional meaning of the word is really Pro-Abortion and is inconsistent with the tenets of Christianity and the teachings in the Bible (Leviticus 24:17).

So, does Planned Parenthood target minorities, as has been claimed? According to CDC data for 2021, 41.5% of women having an abortion were Non-Hispanic Black, 36.3% were Non-Hispanic White, 14.2% were Hispanic with 3.9% of all other races.[17] In 2010, Census Bureau data found that 79% of Planned Parenthood's abortion facilities are located near either Black or Hispanic neighborhoods and 45% are within walking distance of both types of minority neighborhoods.[18]

Black Dignity, an online media resource, boldly states, "We are being targeted. . . . We are 3X more likely to be aborted with almost as many African-American children being aborted as are born. . . . Abortion has reduced our population by 25%. Since 1973 twice as many African-Americans have died from abortion than have died from AIDS, accidents, violent crimes, cancer and heart disease combined."[19]

Planned Parenthood indoctrinates children from kindergarten on up with its own brand of sex education. Children as young as five are now being taught that there are more than two genders. Planned Parenthood is the nation's largest provider of sex education reaching 1.2 million individuals each year, most of whom are in middle school and high school.[20]

"Planned Parenthood does not prioritize the emotional, physical, or spiritual well-being of its clients," Ciancio also writes. " . . . Sex is a gift, and we must not abuse this gift or give this gift to just any random person. . . . This is what our children should understand about sexuality, but Planned Parenthood teaches the opposite. Is this the organization you want teaching your kids about sexuality"?[21]

Life is a precious gift that can only come from our Creator. Treating life in any other manner is not only a crime against humanity but the ultimate travesty and insult to GOD. I find it interesting that abortion advocates have already been born. They would not be here had their mothers exercised their "choice"!

*"For you formed my inward parts; you knitted me together in my mother's womb. I praise you, for I am fearfully and wonderfully made. Wonderful are your works;*

*my soul knows it very well." Psalm 139: 13-14 ESV*

*"Then the LORD God formed the man of dust from the ground and breathed into his nostrils the breath of Life, and the man became a living creature." Genesis 2:7 ESV*

*"But Jesus called them to him, saying, 'Let the children come to me, and do not hinder them, for to such belongs the kingdom of God.'" Luke 18:16 ESV*

*"Whoever takes a human life shall surely be put to death." Leviticus 24:17 ESV*

# Epilogue

JESUS DIED ON THE CROSS FOR ALL OUR SINS, including abortion. We are saved by grace alone. Life is precious and fragile. I believe that our purpose on earth is to do God's will for us and that includes making a difference in others' lives.

Jennifer O'Neill has authored several books detailing many of the topics discussed in this book. I encourage you to research them. They include:

1. *You're Not Alone*: Specifically addresses the post-abortive healing.
2. *From Fallen to Forgiven*: Addresses forgiveness.
3. *Surviving Myself*: Her autobiography.

"May the Lord bless you and keep you; May the Lord make his face shine upon you and be gracious to you; May the Lord lift up His countenance upon you and give you peace." (Numbers 6:24-26 ESV)

# Notes

Chapter 3

1    Davenport Hooker, "Early Fetal Activity in Mammals," *The Yale Journal of Biology and Medicine*, July 1936; and Robert Rules and Landrum B. Shettles, *From Conception to Birth: The Drama of Life's Beginnings* (HarperCollins: New York, 1971).


Appendix C

2    Pew Research Center Short Reads, March 25, 2024.

3    MM Holmes, HS Resnick, DG Kilpatrick, and CL Best, "Rape-Related Pregnancy: Estimates and Descriptive Characteristics from a National Sample of

Women," *Green Journal (The American Journal of Obstetrics and Gynecology)*, August 1996.

4    Pew Research Center Short Reads, March 25, 2024.

5    Pew Research Center Short Reads, March 25, 2024.

6    Pew Research Center Short Reads, March 25, 2024.

7    Pew Research Center Short Reads, March 25, 2024.

8    Pew Research Center Short Reads, March 25, 2024.

Appendix D

9    Susan Ciancio, "12 Planned Parenthood Facts You Should Know," *Human Life International*, April 28, 2023, https://www.hli.org/resources/planned-parenthood-facts/

10    Ciancio, "12 Planned Parenthood Facts You Should Know."

11    "A Force for Hope: Planned Parenthood Annual Report 2023-2024."

12    "A Force for Hope: Planned Parenthood Annual Report 2023-2024."

13   Alina Salganicoff, Laurie Sobel, Ivette Gomez, and Amrutha Ramaswamy, "The Hyde Amendment and Coverage for Abortion Services Under Medicaid in the Post-Roe Era," KFF, March 14, 2024, https://www.kff.org/womens-health-policy/the-hyde-amendment-and-coverage-for-abortion-services-under-medicaid-in-the-post-roe-era

14   Jeannene Maxon, "Funding the Nation's Largest Abortion Business: An Investigation into Public & Private Support of Planned Parenthood," The Lozier Institute, May 12, 2025, https://lozierinstitute.org/funding-the-nations-largest-abortion-provider-an-investigation-into-public-and-private-support-of-planned-parenthood/

15   Brian Clowes, "The Negative Effects of the Pill (and Its Ineffectiveness)," *Human Life International*, May 11, 2017, https://www.hli.org/resources/negative-effects-of-the-pill/

16   MM Holmes, HS Resnick, DG Kilpatrick, and CL Best, "Rape-Related Pregnancy Estimates and Descriptive Characteristics From a National Sample of

Women," *American Journal of Obstetrics & Gynecology*, August 1996, https://pubmed.ncbi.nlm.nih.gov/8765248/

17 Centers for Disease Control and Prevention (CDC): "Abortion Surveillance-United States, 2021," and Guttmacher Institute: "Induced Abortion in the United States."

18 Susan Enouen, "New Research Shows Planned Parenthood Targets Minority Neighborhoods," Life Issues Institute, October 1, 2012, https://lifeissues.org/wp-content/uploads/2023/11/New-Research-PP-Sue-Enouen-pdf.pdf

19 Blackdignity.org

20 Ciancio, "12 Planned Parenthood Facts You Should Know."

21 Ciancio.

# About the Author

D R. JACK MONACO GRADUATED from the Bowman Gray School of Medicine (now Wake Forest School of Medicine) in 1978. He completed a residency in Obstetrics & Gynecology at the St. Francis Hospital & Medical Center in Hartford, Connecticut, where he practiced until his retirement from OB/GYN in 2007. Following his retirement, he transitioned into Integrative and Functional Medicine specializing in hormone restoration for men and women. He is board certified by the American Board of Obstetrics and Gynecology and the American Board of Anti-Aging, Functional and Regenerative Medicine (A4M). During his career, he delivered thousands of babies and was a laparoscopic surgeon specializing in minimally invasive surgery.

After completing his board certification in Anti-aging & Functional Medicine, he founded The Monaco Center for Health & Healing in Glastonbury, Connecticut in 2007. In 2014 he relocated his functional medicine practice to Nashville, Tennessee, where he established the Nashville Hormone & Integrative Medicine Center, which he currently operates. He was a member of the faculty of the American Academy of Anti-Aging and Regenerative Medicine and an oral board examiner. He has lectured throughout the United States, Canada and the Middle East on topics related to hormone restoration, thyroid and adrenal dysfunction and autoimmunity.

Dr. Monaco currently resides in Franklin, Tennessee, with his wife, Lesa, who is also his nurse and partner in healthcare.

Hopefully, this book will cause those women who are uncertain about what to do with their pregnancy to get more information about their options. Consider this book to be the "first step" in your decision and the necessary stimulus for the reader to get an ultrasound. Seeing that life inside of you will certainly have a profound impact on your decision.